D0743834
3 0600 00605 2477

FARMING: A HAND BOOK

Books by Wendell Berry

Novels

Nathan Coulter

A Place on Earth

The Memory of Old Jack

Poetry

The Broken Ground

Openings

Findings

Farming: A Hand Book

The Country of Marriage

Clearing

Essays

The Long-Legged House

The Hidden Wound

The Unforeseen Wilderness

A Continuous Harmony

WENDELL BERRY

Farming: A Hand Book

A HARVEST/HBJ BOOK
HARCOURT BRACE JOVANOVICH, PUBLISHERS
SAN DIEGO NEW YORK LONDON

Library of Congress Cataloging in Publication Data
Berry, Wendell, 1934–
Farming : a handbook.
"A Harvest/HBJ book."
Bibliography: p.
I. Title.
PS3552.E75F3 1985 811'.54 84-22555
ISBN 0-15-630171-7

Some of the poems in this volume previously appeared in *Apple, Chelsea, Colloquy, Field, The Florida Quarterly, Free You, The Hudson Review, The Iowa Review, Kayak, Lillabulero, Mademoiselle, Mill Mountain Review, Monk's Pond, Mountain Life & Work, The Nation, The New York Times, The Penny Paper, The Poetry Bag, Sequoia Magazine, south, Southern Poetry Review, Sou'wester Literary Magazine, Stand, Sumac, Tennessee Poetry Journal, The Vassar Review.* "September 2, 1969" appeared originally in *Poetry.*

Printed in the United States of America

FG

For Owen and Loyce

CONTENTS

I

II

III

IV

I

THE MAN BORN TO FARMING

The grower of trees, the gardener, the man born to farming,
whose hands reach into the ground and sprout,
to him the soil is a divine drug. He enters into death
yearly, and comes back rejoicing. He has seen the light lie
down
in the dung heap, and rise again in the corn.
His thought passes along the row ends like a mole.
What miraculous seed has he swallowed
that the unending sentence of his love flows out of his mouth
like a vine clinging in the sunlight, and like water
descending in the dark?

THE STONES

I owned a slope full of stones.
Like buried pianos they lay in the ground,
shards of old sea-ledges, stumbling blocks
where the earth caught and kept them
dark, an old music mute in them
that my head keeps now I have dug them out.
I broke them where they slugged in their dark
cells, and lifted them up in pieces.
As I piled them in the light
I began their music. I heard their old lime
rouse in breath of song that has not left me.
I gave pain and weariness to their bearing out.
What bond have I made with the earth,
having worn myself against it? It is a fatal singing
I have carried with me out of that day.
The stones have given me music
that figures for me their holes in the earth
and their long lying in them dark.
They have taught me the weariness that loves the ground,
and I must prepare a fitting silence.

THE SUPPLANTING

Where the road came, no longer bearing men,
but briars, honeysuckle, buckbush and wild grape,
the house fell to ruin, and only the old wife's daffodils
rose in spring among the wild vines to be domestic
and to keep the faith, and her peonies drenched the tangle
with white bloom. For a while in the years of its wilderness
a wayfaring drunk slept clinched to the floor there
in the cold nights. And then I came, and set fire
to the remnants of house and shed, and let time
hurry in the flame. I fired it so that all
would burn, and watched the blaze settle on the waste
like a shawl. I knew those old ones departed
then, and I arrived. As the fire fed, I felt rise in me
something that would not bear my name—something that
 bears us
through the flame, and is lightened of us, and is glad.

SOWING

In the stilled place that once was a road going down
from the town to the river, and where the lives of marriages
grew
a house, cistern and barn, flowers, the tilted stone of borders,
and the deeds of their lives ran to neglect, and honeysuckle
and then the fire overgrew it all, I walk heavy
with seed, spreading on the cleared hill the beginnings
of green, clover and grass to be pasture. Between
history's death upon the place and the trees that would have
come
I claim, and act, and am mingled in the fate of the world.

THE FAMILIAR

The hand is risen from the earth,
the sap risen, leaf come back to branch,
bird to nest crotch. Beans lift
their heads up in the row. The known
returns to be known again. Going
and coming back, it forms its curves,
a nerved ghostly anatomy in the air.

THE FARMER AMONG THE TOMBS

I am oppressed by all the room taken up by the dead,
their headstones standing shoulder to shoulder,
the bones imprisoned under them.
Plow up the graveyards! Haul off the monuments!
Pry open the vaults and the coffins
so the dead may nourish their graves
and go free, their acres traversed all summer
by crop rows and cattle and foraging bees.

FOR THE REBUILDING OF A HOUSE

To know the inhabiting reasons
of trees and streams, old men
who shed their lives
on the world like leaves,
I watch them go.
And I go. I build
the place of my leaving.

The days arc into vision
like fish leaping, their shining
caught in the stream.
I watch them go
in homage and sorrow.
I build the place of my dream.
I build the place of my leaving
that the dark may come clean.

THE SPRINGS

In a country without saints or shrines
I knew one who made his pilgrimage
to springs, where in his life's dry years
his mind held on. Everlasting,
people called them, and gave them names.
The water broke into sounds and shinings
at the vein mouth, bearing the taste
of the place, the deep rock, sweetness
out of the dark. He bent and drank
in bondage to the ground.

WATER

I was born in a drouth year. That summer
my mother waited in the house, enclosed
in the sun and the dry ceaseless wind,
for the men to come back in the evenings,
bringing water from a distant spring.
Veins of leaves ran dry, roots shrank.
And all my life I have dreaded the return
of that year, sure that it still is
somewhere, like a dead enemy's soul. Fear
of dust in my mouth is always with me,
and I am the faithful husband of the rain,
I love the water of wells and springs
and the taste of roofs in the water of cisterns.
I am a dry man whose thirst is praise
of clouds, and whose mind is something of a cup.
My sweetness is to wake in the night
after days of dry heat, hearing the rain.

RAIN

It is a day of the earth's renewing without any man's doing
    or help.
Though I have fields I do not go out to work in them.
Though I have crops standing in rows I do not go out
to look at them or gather what has ripened or hoe the weeds
    from the balks.
Though I have animals I stay dry in the house while they
    graze in the wet.
Though I have buildings they stand closed under their roofs.
Though I have fences they go without me.
My life stands in place, covered, like a hay rick or a
    mushroom.

SLEEP

I love to lie down weary
under the stalk of sleep
growing slowly out of my head,
the dark leaves meshing.

TO KNOW THE DARK

To go in the dark with a light is to know the light.
To know the dark, go dark. Go without sight,
and find that the dark, too, blooms and sings,
and is traveled by dark feet and dark wings.

WINTER NIGHT POEM FOR MARY

As I started home after dark
I looked into the sky and saw the new moon,
an old man with a basket on his arm.
He walked among the cedars in the bare woods.
They stood like guardians, dark
as he passed. He might have been singing,
or he might not. He might have been sowing
the spring flowers, or he might not. But I saw him
with his basket, going along the hilltop.

WINTER NIGHTFALL

The fowls speak and sing, settling for the night.
The mare shifts in the bedding.
In her womb her foal sleeps and grows,
within and within and within. Her jaw grinds,
meditative in the fragrance of timothy.
Soon now my own rest will come.
The silent river flows on in the dusk, miles and miles.
Outside the walls and on the roof and in the woods
the cold rain falls.

FEBRUARY 2, 1968

In the dark of the moon, in flying snow, in the dead of winter,
war spreading, families dying, the world in danger,
I walk the rocky hillside, sowing clover.

MARCH 22, 1968

As spring begins the river rises,
filling like the sorrow of nations
—uprooted trees, soil of squandered mountains,
the debris of kitchens, all passing
seaward. At dawn snow began to fall.
The ducks, moving north, pass
like shadows through the falling white.
The jonquils, half open, bend down with its weight.
The plow freezes in the furrow.
In the night I lay awake, thinking
of the river rising, the spring heavy
with official meaningless deaths.

THE MORNING'S NEWS

To moralize the state, they drag out a man,
and bind his hands, and darken his eyes
with a black rag to be free of the light in them,
and tie him to a post, and kill him.
And I am sickened by complicity in my race.
To kill in hot savagery like a beast
is understandable. It is forgivable and curable.
But to kill by design, deliberately, without wrath,
that is the sullen labor that perfects Hell.
The serpent is gentle, compared to man.
It is man, the inventor of cold violence,
death as waste, who has made himself lonely
among the creatures, and set himself aside
from creation, so that he cannot labor
in the light of the sun with hope,
or sit at peace in the shade of any tree.
The morning's news drives sleep out of the head
at night. Uselessness and horror hold the eyes
open to the dark. Weary, we lie awake
in the agony of the old giving birth to the new
without assurance that the new will be better.
I look at my son, whose eyes are like a young god's,
they are so open to the world.
I look at my sloping fields now turning
green with the young grass of April. What must I do
to go free? I think I must put on
a deathlier knowledge, and prepare to die
rather than enter into the design of man's hate.

I will purge my mind of the airy claims
of church and state, and observe the ancient wisdom
of tribesman and peasant, who understood
they labored on the earth only to lie down in it
in peace, and were content. I will serve the earth
and not pretend my life could be better served.
My life is only the earth risen up
a little way into the light, among the leaves.
Another morning comes with its strange cure.
The earth is news. Though the river floods
and the spring is cold, my heart goes on,
faithful to a mystery in a cloud,
and the summer's garden continues its descent
through me, toward the ground.

ENRICHING THE EARTH

To enrich the earth I have sowed clover and grass
to grow and die. I have plowed in the seeds
of winter grains and of various legumes,
their growth to be plowed in to enrich the earth.
I have stirred into the ground the offal
and the decay of the growth of past seasons
and so mended the earth and made its yield increase.
All this serves the dark. Against the shadow
of veiled possibility my workdays stand
in a most asking light. I am slowly falling
into the fund of things. And yet to serve the earth,
not knowing what I serve, gives a wideness
and a delight to the air, and my days
do not wholly pass. It is the mind's service,
for when the will fails so do the hands
and one lives at the expense of life.
After death, willing or not, the body serves,
entering the earth. And so what was heaviest
and most mute is at last raised up into song.

A WET TIME

The land is an ark, full of things waiting.
Underfoot it goes temporary and soft, tracks
filling with water as the foot is raised.
The fields, sodden, go free of plans. Hands
become obscure in their use, prehistoric.
The mind passes over changed surfaces
like a boat, drawn to the thought of roofs
and to the thought of swimming and wading birds.
Along the river croplands and gardens
are buried in the flood, airy places grown dark
and silent beneath it. Under the slender branch
holding the new nest of the hummingbird
the river flows heavy with earth, the water
turned the color of broken slopes. I stand
deep in the mud of the shore, like a stake
planted to measure the rise, the water rising,
the earth falling to meet it. A great cottonwood
passes down, the leaves shivering as the roots
drag the bottom. I turn like an ancient worshipper
to the thought of solid ground. I was not ready for this
parting, my native land putting out to sea.

THE SILENCE

What must a man do to be at home in the world?
There must be times when he is here
as though absent, gone beyond words into the woven
    shadows
of the grass and the flighty darknesses
of leaves shaking in the wind, and beyond
the sense of the weariness of engines and of his own heart,
his wrongs grown old unforgiven. It must be with him
as though his bones fade beyond thought
into the shadows that grow out of the ground
so that the furrow he opens in the earth opens
in his bones, and he hears the silence
of the tongues of the dead tribesmen buried here
a thousand years ago. And then what presences will rise up
before him, weeds bearing flowers, and the dry wind
rain! What songs he will hear!

IN THIS WORLD

The hill pasture, an open place among the trees,
tilts into the valley. The clovers and tall grasses
are in bloom. Along the foot of the hill
dark floodwater moves down the river.
The sun sets. Ahead of nightfall the birds sing.
I have climbed up to water the horses
and now sit and rest, high on the hillside,
letting the day gather and pass. Below me
cattle graze out across the wide fields of the bottomlands,
slow and preoccupied as stars. In this world
men are making plans, wearing themselves out,
spending their lives, in order to kill each other.

THE NEW ROOF

On the housetop, the floor of the boundless
where birds and storms fly and disappear,
and the valley opened over our heads, a leap
of clarity between the hills, we bent five days
in the sun, tearing free the old roof, nailing on
the new, letting the sun touch for once
in fifty years the dusky rafters, and then
securing the house again in its shelter and shade.
Thus like a little ledge a piece of my history
has come between me and the sky.

A PRAISE

His memories lived in the place
like fingers locked in the rock ledges
like roots. When he died
and his influence entered the air
I said, Let my mind be the earth
of his thought, let his kindness
go ahead of me. Though I do not escape
the history barbed in my flesh,
certain wise movements of his hands,
the turns of his speech
keep with me. His hope of peace
keeps with me in harsh days,
the shell of his breath dimming away
three summers in the earth.

ON THE HILL LATE AT NIGHT

The ripe grassheads bend in the starlight
in the soft wind, beneath them the darkness
of the grass, fathomless, the long blades
rising out of the well of time. Cars
travel the valley roads below me, their lights
finding the dark, and racing on. Above
their roar is a silence I have suddenly heard,
and felt the country turn under the stars
toward dawn. I am wholly willing to be here
between the bright silent thousands of stars
and the life of the grass pouring out of the ground.
The hill has grown to me like a foot.
Until I lift the earth I cannot move.

THE BARN

While we unloaded the hay from the truck, building
the great somnolence of the ricked bales, the weather
kept up its movement over us, the rain dashed and drove
against the roof, and in the close heat we sweated
to the end of the load. The fresh warm sweet smell
of new timothy in it, the barn is a nut ripened
in forethought of cold. Weighted now, it turns
toward the future generously, spacious
in its intent, the fledged young of the barn swallows
fluttering on the rim of the nest, the brown bats
hanging asleep, folded, beneath the rafters.
And we rest, having done what men are best at.

THE BUILDINGS

The buildings are all womanly. Their roofs
are like the flanks of mares, the arms and the hair of wives.
The future prepares its satisfaction in them.
In their dark heat I labor all summer, making them ready.
A time of death is coming, and they desire to live.
It is only the labor surrounding them that is manly,
the seasonal bringing in from the womanly fields
to the womanly enclosures. The house too yearns for life,
and hot paths come to it out of the garden and the fields,
full of the sun and weary. The wifeliness of my wife
is its welcome, a vine with yellow flowers shading the door.

THE SEEDS

The seeds begin abstract as their species,
remote as the name on the sack
they are carried home in: Fayette Seed Company
Corner of Vine and Rose. But the sower
going forth to sow sets foot
into time to come, the seeds falling
on his own place. He has prepared a way
for his life to come to him, if it will.
Like a tree, he has given roots
to the earth, and stands free.

THE WISH TO BE GENEROUS

All that I serve will die, all my delights,
the flesh kindled from my flesh, garden and field,
the silent lilies standing in the woods,
the woods and the hill and the whole earth, all
will burn in man's evil, or dwindle
in its own age. Let the world bring on me
the sleep of darkness without stars, so I may know
my little light taken from me into the seed
of the beginning and the end, so I may bow
to mystery, and take my stand on the earth
like a tree in a field, passing without haste
or regret toward what will be, my life
a patient willing descent into the grass.

AIR AND FIRE

From my wife and household and fields
that I have so carefully come to in my time
I enter the craziness of travel,
the reckless elements of air and fire.
Having risen up from my native land,
I find myself smiled at by beautiful women,
making me long for a whole life
to devote to each one, making love to her
in some house, in some way of sleeping
and waking I would make only for her.
And all over the country I find myself
falling in love with houses, woods, and farms
that I will never set foot in.
My eyes go wandering through America,
two wayfaring brothers, resting in silence
against the forbidden gates. O what if
an angel came to me, and said,
"Go free of what you have done. Take
what you want." The atoms of the blood
and brain and bone strain apart
at the thought. What I am is the way home.
Like rest after a sleepless night
my old love comes on me in midair.

THE LILIES

Amid the gray trunks of ancient trees we found
the gay woodland lilies nodding on their stems,
frail and fair, so delicately balanced the air
held or moved them as it stood or moved.
The ground that slept beneath us woke in them
and made a music of the light, as it had waked
and sung in fragile things unnumbered years,
and left their kind no less symmetrical and fair
for all that time. Does my land have the health
of this, where nothing falls but into life?

INDEPENDENCE DAY

for Gene Meatyard

Between painting a roof yesterday and the hay
harvest tomorrow, a holiday in the woods
under the grooved trunks and branches, the roof
of leaves lighted and shadowed by the sky.
As America from England, the woods stands free
from politics and anthems. So in the woods I stand
free, knowing my land. My country, tis of the
drying pools along Camp Branch I sing
where the water striders walk like Christ,
all sons of God, and of the woods grown old
on the stony hill where the thrush's song rises
in the light like a curling vine and the bobwhite's
whistle opens in the air, broad and pointed like a leaf.

II

I bid you to a one-man revolution—
The only revolution that is coming.

*

We're too unseparate. And going home
From company means coming to our senses.

—ROBERT FROST, "Build Soil"

A STANDING GROUND

Flee fro the prees, and dwelle with sothfastnesse;
Suffyce unto thy thyng, though hit be smal . . .

However just and anxious I have been,
I will stop and step back
from the crowd of those who may agree
with what I say, and be apart.
There is no earthly promise of life or peace
but where the roots branch and weave
their patient silent passages in the dark;
uprooted, I have been furious without an aim.
I am not bound for any public place,
but for ground of my own
where I have planted vines and orchard trees,
and in the heat of the day climbed up
into the healing shadow of the woods.
Better than any argument is to rise at dawn
and pick dew-wet red berries in a cup.

SONG IN A YEAR OF CATASTROPHE

I began to be followed by a voice saying:
"It can't last. It can't last.
Harden yourself. Harden yourself.
Be ready. Be ready."

"Go look under the leaves,"
it said, "for what is living there
is long dead in your tongue."
And it said, "Put your hands
into the earth. Live close
to the ground. Learn the darkness.
Gather round you all
the things that you love, name
their names, prepare
to lose them. It will be
as if all you know were turned
around within your body."

And I went and put my hands
into the ground, and they took root
and grew into a season's harvest.
I looked behind the veil
of the leaves, and heard voices
that I knew had been dead
in my tongue years before my birth.
I learned the dark.

And still the voice stayed with me.
Waking in the early mornings,
I could hear it, like a bird
bemused among the leaves,
a mockingbird idly singing
in the autumn of catastrophe:
"Be ready. Be ready.
Harden yourself. Harden yourself."

And I heard the sound
of a great engine pounding
in the air, and a voice asking:
"Change or slavery?
Hardship or slavery?"
and voices answering:
"Slavery! Slavery!"
And I was afraid, loving
what I knew would be lost.

Then the voice following me said:
"You have not yet come close enough.
Come nearer the ground. Learn
from the woodcock in the woods
whose feathering is a ritual
of the fallen leaves,
and from the nesting quail
whose speckling makes her hard to see
in the long grass.
Study the coat of the mole.

For the farmer shall wear
the greenery and the furrows
of his fields, and bear
the long standing of the woods."

And I asked: "You mean a death, then?"
"Yes," the voice said. "Die
into what the earth requires of you."
Then I let go all holds, and sank
like a hopeless swimmer into the earth,
and at last came fully into the ease
and the joy of that place,
all my lost ones returning.

9/28/68

THE CURRENT

Having once put his hand into the ground,
seeding there what he hopes will outlast him,
a man has made a marriage with his place,
and if he leaves it his flesh will ache to go back.
His hand has given up its birdlife in the air.
It has reached into the dark like a root
and begun to wake, quick and mortal, in timelessness,
a flickering sap coursing upward into his head
so that he sees the old tribespeople bend
in the sun, digging with sticks, the forest opening
to receive their hills of corn, squash, and beans,
their lodges and graves, and closing again.
He is made their descendant, what they left
in the earth rising into him like a seasonal juice.
And he sees the bearers of his own blood arriving,
the forest burrowing into the earth as they come,
their hands gathering the stones up into walls,
and relaxing, the stones crawling back into the ground
to lie still under the black wheels of machines.
The current flowing to him through the earth
flows past him, and he sees one descended from him,
a young man who has reached into the ground,
his hand held in the dark as by a hand.

THE MAD FARMER REVOLUTION

BEING A FRAGMENT
OF THE NATURAL HISTORY OF NEW EDEN,
IN HOMAGE
TO MR. ED McCLANAHAN, ONE OF THE LOCALS

The mad farmer, the thirsty one,
went dry. When he had time
he threw a visionary high
lonesome on the holy communion wine.
"It is an awesome event
when an earthen man has drunk
his fill of the blood of a god,"
people said, and got out of his way.
He plowed the churchyard, the
minister's wife, three graveyards
and a golf course. In a parking lot
he planted a forest of little pines.
He sanctified the groves,
dancing at night in the oak shades
with goddesses. He led
a field of corn to creep up
and tassel like an Indian tribe
on the courthouse lawn. Pumpkins
ran out to the ends of their vines
to follow him. Ripe plums
and peaches reached into his pockets.
Flowers sprang up in his tracks
everywhere he stepped. And then
his planter's eye fell on

that parson's fair fine lady
again. "O holy plowman," cried she,
"I am all grown up in weeds.
Pray, bring me back into good tilth."
He tilled her carefully
and laid her by, and she
did bring forth others of her kind,
and others, and some more.
They sowed and reaped till all
the countryside was filled
with farmers and their brides sowing
and reaping. When they died
they became two spirits of the woods.

On their graves were written
these words without sound:
"Here lies Saint Plowman.
Here lies Saint Fertile Ground."

THE CONTRARINESS OF THE MAD FARMER

I am done with apologies. If contrariness is my
inheritance and destiny, so be it. If it is my mission
to go in at exits and come out at entrances, so be it.
I have planted by the stars in defiance of the experts,
and tilled somewhat by incantation and by singing,
and reaped, as I knew, by luck and Heaven's favor,
in spite of the best advice. If I have been caught
so often laughing at funerals, that was because
I knew the dead were already slipping away,
preparing a comeback, and can I help it?
And if at weddings I have gritted and gnashed
my teeth, it was because I knew where the bridegroom
had sunk his manhood, and knew it would not
be resurrected by a piece of cake. "Dance" they told me,
and I stood still, and while they stood
quiet in line at the gate of the Kingdom, I danced.
"Pray" they said, and I laughed, covering myself
in the earth's brightnesses, and then stole off gray
into the midst of a revel, and prayed like an orphan.
When they said "I know that my Redeemer liveth,"
I told them "He's dead." And when they told me
"God is dead," I answered "He goes fishing every day
in the Kentucky River. I see Him often."
When they asked me would I like to contribute
I said no, and when they had collected
more than they needed, I gave them as much as I had.
When they asked me to join them I wouldn't,

and then went off by myself and did more
than they would have asked. "Well, then" they said
"go and organize the International Brotherhood
of Contraries," and I said "Did you finish killing
everybody who was against peace?" So be it.
Going against men, I have heard at times a deep harmony
thrumming in the mixture, and when they ask me what
I say I don't know. It is not the only or the easiest
way to come to the truth. It is one way.

THE FARMER AND THE SEA

The sea always arriving,
hissing in pebbles, is breaking
its edge where the landsman
squats on his rock. The dark
of the earth is familiar to him,
close mystery of his source
and end, always flowering
in the light and always
fading. But the dark of the sea
is perfect and strange,
the absence of any place,
immensity on the loose.
Still, he sees it is another
keeper of the land, caretaker,
shaking the earth, breaking it,
clicking the pieces, but somewhere
holding deep fields yet to rise,
shedding its richness on them
silently as snow, keeper and maker
of places wholly dark. And in him
something dark applauds.

EARTH AND FIRE

In this woman the earth speaks.
Her words open in me, cells of light
flashing in my body, and make a song
that I follow toward her out of my need.
The pain I have given her I wear
like another skin, tender, the air
around me flashing with thorns.
And yet such joy as I have given her
sings in me and is part of her song.
The winds of her knees shake me
like a flame. I have risen up from her,
time and again, a new man.

THE MAD FARMER IN THE CITY

". . . a field woman is a portion
of the field; she has somehow lost
her own margin . . ."—THOMAS HARDY

As my first blow against it, I would not stay.
As my second, I learned to live without it.
As my third, I went back one day and saw
that my departure had left a little hole
where some of its strength was flowing out,
and I heard the earth singing beneath the street.
Singing quietly myself, I followed the song
among the traffic. Everywhere I went, singing,
following the song, the stones cracked,
and I heard it stronger. I heard it strongest
in the presence of women. There was one I met
who had the music of the ground in her, and she
was its dancer. "O Exile," I sang, "for want of you
there is a tree that has borne no leaves
and a planting season that will not turn warm."
Looking at her, I felt a tightening of roots
under the pavement, and I turned and went
with her a little way, dancing beside her.
And I saw a black woman still inhabiting
as in a dream the space of the open fields
where she had bent to plant and gather. She stood
rooted in the music I heard, pliant and proud
as a stalk of wheat with the grain heavy. No man
with the city thrusting angles in his brain
is equal to her. To reach her he must tear it down.

Wherever lovely women are the city is undone,
its geometry broken in pieces and lifted,
its streets and corners fading like mist at sunrise
above groves and meadows and planted fields.

THE BIRTH (Near Port William)

They were into the lambing, up late.
Talking and smoking around their lantern,
they squatted in the barn door, left open
so the quiet of the winter night
diminished what they said. The chill
had begun to sink into their clothes.
Now and then they raised their hands
to breathe on them. The youngest one
yawned and shivered.

"Damn," he said,
"I'd like to be asleep. I'd like to be
curled up in a warm nest like an old
groundhog, and sleep till spring."

"When I was your age, Billy, it wasn't
sleep I thought about," Uncle Stanley said.
"Last few years here I've took to sleeping."

And Raymond said: "To sleep till spring
you'd have to have a trust in things
the way animals do. Been a long time,
I reckon, since people felt safe enough
to sleep more than a night. You might
wake up someplace you didn't go to sleep at."

They hushed a while, as if to let the dark
brood on what they had said. Behind them

a sheep stirred in the bedding and coughed.
It was getting close to midnight.
Later they would move back along the row
of penned ewes, making sure the newborn
lambs were well dried, and had sucked,
and then they would go home cold to bed.
The barn stood between the ridgetop
and the woods along the bluff. Below
was the valley floor and the river
they could not see. They could hear
the wind dragging its underside
through the bare branches of the woods.
And suddenly the wind began to carry
a low singing. They looked across
the lantern at each other's eyes
and saw they all had heard. They stood,
their huge shadows rising up around them.
The night had changed. They were already
on their way—dry leaves underfoot
and mud under the leaves—to another barn
on down along the woods' edge,
an old stripping room, where by the light
of the open stove door they saw the man,
and then the woman and the child
lying on a bed of straw on the dirt floor.

"Well, look a there," the old man said.
"First time this ever happened here."

And Billy, looking, and looking away,
said: "Howdy. Howdy. Bad night."

And Raymond said: "There's a first
time, they say, for everything."

And that,
he thought, was as reassuring as anything
was likely to be, and as he needed it to be.
They did what they could. Not much.
They brought a piece of rug and some sacks
to ease the hard bed a little, and one
wedged three dollar bills into a crack
in the wall in a noticeable place.
And they stayed on, looking, looking away,
until finally the man said they were well
enough off, and should be left alone.
They went back to their sheep. For a while
longer they squatted by their lantern
and talked, tired, wanting sleep, yet stirred
by wonder—old Stanley too, though he would not
say so.

"Don't make no difference," he said.
"They'll have 'em anywhere. Looks like a man
would have a right to be born in bed, if not
die there, but he don't."

"But you heard
that singing in the wind," Billy said.
"What about that?"

"Ghosts. They do that way."

"Not that way."

"Scared him, it did."
The old man laughed. "We'll have to hold
his damn hand for him, and lead him home."

"It don't even bother you," Billy said.
"You go right on just the same. But you heard."

"Now that I'm old I sleep in the dark.
That ain't what I used to do in it. I heard
something."

"You heard a good deal more
than you'll understand," Raymond said,
"or him or me either."

They looked at him.
He had, they knew, a talent for unreasonable
belief. He could believe in tomorrow
before it became today—a human enough
failing, and they were tolerant.

He said:
"It's the old ground trying it again.
Solstice, seeding and birth—it never
gets enough. It wants the birth of a man
to bring together sky and earth, like a stalk
of corn. It's not death that makes the dead
rise out of the ground, but something alive

straining up, rooted in darkness, like a vine.
That's what you heard. If you're in the right mind
when it happens, it can come on you strong,
and you might hear music passing on the wind,
or see a light where there wasn't one before."

"Well, how do you know if it amounts to anything?"

"You don't. It usually don't. It would take
a long long time to ever know."

But that night
and other nights afterwards, up late,
there was a feeling in them—familiar
to them, but always startling in its strength—
like the thought, on a winter night,
of the lambing ewes dry-bedded and fed,
and the thought of the wild creatures warm
asleep in their nests, deep underground.

AWAKE AT NIGHT

Late in the night I pay
the unrest I owe
to the life that has never lived
and cannot live now.
What the world could be
is my good dream
and my agony when, dreaming it,
I lie awake and turn
and look into the dark.
I think of a luxury
in the sturdiness and grace
of necessary things, not
in frivolity. That would heal
the earth, and heal men.
But the end, too, is part
of the pattern, the last
labor of the heart:
to learn to lie still,
one with the earth
again, and let the world go.

PRAYERS AND SAYINGS OF THE MAD FARMER

for James Baker Hall

I

It is presumptuous and irresponsible to pray for other people. A good man would pray only for himself—that he have as much good as he deserves, that he not receive more good or more evil than he deserves, that he bother nobody, that he not be bothered, that he want less. Praying thus for himself, he should prepare to live with the consequences.

II

At night make me one with the darkness.
In the morning make me one with the light.

III

If a man finds it neccessary to eat garbage, he should resist the temptation to call it a delicacy.

IV

Don't pray for the rain to stop.
Pray for good luck fishing
when the river floods.

V

Don't own so much clutter that you will be relieved to see your house catch fire.

VI

Beware of the machinery of longevity. When a man's life is

over the decent thing is for him to die. The forest does not withhold itself from death. What it gives up it takes back.

VII

Put your hands into the mire.
They will learn the kinship
of the shaped and the unshapen,
the living and the dead.

VIII

When I rise up
let me rise up joyful
like a bird.

When I fall
let me fall without regret
like a leaf.

IX

Sowing the seed,
my hand is one with the earth.

Wanting the seed to grow,
my mind is one with the light.

Hoeing the crop,
my hands are one with the rain.

Having cared for the plants,
my mind is one with the air.

Hungry and trusting,
my mind is one with the earth.

Eating the fruit,
my body is one with the earth.

X

Let my marriage be brought to the ground.
Let my love for this woman enrich the earth.

What is its happiness but preparing its place?
What is its monument but a rich field?

XI

By the excellence of his work the workman is a neighbor. By selling only what he would not despise to own the salesman is a neighbor. By selling what is good his character survives his market.

XII

Let me wake in the night
and hear it raining
and go back to sleep.

XIII

Don't worry and fret about the crops. After you have done all you can for them, let them stand in the weather on their own.

If the crop of any one year was all, a man would have to cut his throat every time it hailed.

But the *real* products of any year's work are the farmer's mind and the cropland itself.

If he raises a good crop at the cost of belittling himself and diminishing the ground, then he has gained nothing. He will have to begin all over again the next spring, worse off than before.

Let him receive the season's increment into his mind. Let him work it into the soil.

The finest growth that farmland can produce is a careful farmer.

Make the human race a better head. Make the world a better piece of ground.

THE SATISFACTIONS OF THE MAD FARMER

Growing weather; enough rain;
the cow's udder tight with milk;
the peach tree bent with its yield;
honey golden in the white comb;

the pastures deep in clover and grass,
enough, and more than enough;

the ground, new worked, moist
and yielding underfoot, the feet
comfortable in it as roots;

the early garden: potatoes, onions,
peas, lettuce, spinach, cabbage, carrots,
radishes, marking their straight rows
with green, before the trees are leafed;

raspberries ripe and heavy amid their foliage,
currants shining red in clusters amid their foliage,
strawberries red ripe with the white
flowers still on the vines—picked
with the dew on them, before breakfast;

grape clusters heavy under broad leaves,
powdery bloom on fruit black with sweetness
—an ancient delight, delighting;

the bodies of children, joyful
without dread of their spending,
surprised at nightfall to be weary;

the bodies of women in loose cotton,
cool and closed in the evenings
of summer, like contented houses;

the bodies of men, competent in the heat
and sweat and weight and length
of the day's work, eager in their spending,
attending to nightfall, the bodies of women;

sleep after love, dreaming
white lilies blooming
coolly out of my flesh;

after sleep, the sense of being enabled
to go on with work, morning a clear gift;

the maidenhood of the day,
cobwebs unbroken in the dewy grass;

the work of feeding and clothing and housing,
done with more than enough knowledge
and with more than enough love,
by men who do not have to be told;

any building well built, the rafters
firm to the walls, the walls firm,
the joists without give,

the proportions clear,
the fitting exact, even unseen,
bolts and hinges that turn home
without a jiggle;

any work worthy
of the day's maidenhood;

any man whose words
lead precisely to what exists,
who never stoops to persuasion;

the talk of friends, lightened and cleared
by all that can be assumed;

deer tracks in the wet path,
the deer sprung from them, gone on;

live streams, live shiftings
of the sun in the summer woods;

the great hollow-trunked beech,
a landmark I loved to return to,
its leaves gold-lit on the silver
branches in the fall: blown down
after a hundred years of standing,
a footbridge over the stream;

the quiet in the woods of a summer morning,
the voice of a pewee passing through it
like a tight silver wire;

a little clearing among cedars,
white clover and wild strawberries
beneath an opening to the sky
—heavenly, I thought it,
so perfect; had I foreseen it
I would have desired it
no less than it deserves;

fox tracks in snow, the impact
of lightness upon lightness,
unendingly silent.

What I know of spirit is astir
in the world. The god I have always expected
to appear at the woods' edge, beckoning,
I have always expected to be
a great relisher of the world, its good
grown immortal in his mind.

III

THE BRINGER OF WATER

The people in the play:

Mat Feltner

A farmer of the town of Port William, Kentucky. He is now sixty-five years old.

Margaret Feltner

Mat's wife, about sixty-three.

Hannah Feltner

The widow of Virgil, the son of Mat and Margaret, killed in the final spring of World War II.

Little Margaret

Daughter of Virgil and Hannah Feltner. She is three years old.

Henry Catlett

Son of Mat's and Margaret's daughter.

Old Jack Beechum

Mat's uncle, a farmer, now eighty-four years old, living in town.

An old man, an old woman, a young woman

Neighbors of the Feltners.

Burley and Jarrat Coulter

Brothers. Farmers. Burley is fifty-three, Jarrat fifty-eight.

Nathan Coulter

Jarrat's son, twenty-four years old, a veteran of the war.

SCENE 1

The Feltners' back porch in early July, 1948. It is after the noon meal, for Mat a moment of deliberate ease before the afternoon's work in the field. Mat and Margaret sit together on the porch. Though their chairs are rocking chairs, they sit still. An earthen jug of water stands at Mat's feet. Hannah sits in a swing hung from the limb of a shade tree out in the yard. Sewing, she looks up from time to time to watch her daughter and Henry Catlett, who sits on the grass near her. To amuse the little girl Henry is building a house of brightly colored blocks.

Margaret: She worries me. Three years
now since Virgil was killed
in the war, and still she keeps
to herself, saying nothing
of what she's concerned with most.

Mat: If it wasn't for the children
—and Uncle Jack, whatever
company he is to her—
she might as well be alone.
She troubles me too.
I'd like to see her happy.
But mostly I'd like to see her
turn back to the world
and take up life again
for what it's worth.

That she seems so fearful
is what worries me.

Margaret: Well, I know the fear
of change, how time passing
threatens her with the worst
she can think of. The past
taints the future. Sometimes
so much is lost in it
that only enough seems left
to grieve and to be afraid.
But you're right. The fear
of nothing in particular
is a waste.

Mat: I reckon after all
it's lucky she has the child.

Margaret: It's not in her children
that a woman can live. Not
in the future. She lives
in living with her man
who is the present, asking,
changing, filling hand
and heart, not with the hoped
or the mourned, but with his life
as it is, what is possible in time.

Mat: Maybe so. A nice thing anyhow,
mam, for a lady to say
to her husband. It may be

a change will come. I've noticed
Nathan Coulter looking at her
when she brings water to the field
or he comes by the house.
It's not the look, I'd say,
of a lover who has been at all
encouraged, but it's a look
I understand.

Margaret: And it's a look
that Hannah understands,
don't doubt it—and even
values, as a young woman has to.
But that it moves her warns her
to hold herself away,
and she treats him with a cold
politeness women have, that won't
speak until it has to.

They sit in silence a moment. And then Mat stands up stiffly, a little wearily.

Mat: Oh me! I reckon I'd better
be getting on back.
They'll already be there.

He picks up his jug and starts across the back yard toward a little slatted gate. Hannah looks up at him as he passes, smiling at him.

Hannah: I'm sorry you have to go
back out there. It's so hot.

Mat: Yes. It's hot. It's beating down.

Hannah: You'll be needing water
later?

Mat: About three o'clock,
I expect. A fresh drink then
would help us last till night.

He goes on across the yard and through the gate. The others watch him go.

SCENE 2

Hannah, accompanied by Henry and Old Jack, is on the way to the field to take water to the men at work there. Generally, Henry goes in front of Hannah, adventuring, calling back. Old Jack comes last, usually too far behind to talk to the others. From time to time, glancing back, Hannah stops to let the old man rest, though to spare his feelings she always makes it appear that she wants to rest. At the beginning they come up to and pass a man and two women picking pole beans in a neighboring garden.

Hannah: I became the vessel of all
of him that could live,
his seed's earth. I held
still as if balanced
in a light little boat, for fear
I'd hurt the last chance
his life had to be alive.
Then little Margaret was born.
There could be no other
like her, and still I kept
still, thinking all the time
of her. And now she's growing,
not a baby any more.
At times I see her draw away
from me toward the others,
and I know her life
is hers, and mine is mine.

I begin to feel again
the claims on me my life has.
As though I felt my body
touched in the night, I want
to be talked to, touched,
for only my own sake.

The old woman:

Here she comes with her bucket
again, with the little Catlett boy
and that scandalous old man
Jack Beechum, them two
she trusts herself to. I'd say
she's a strange one, with her looks
and all, living like a nun.

The old man: Nuns, they say, is married
to Jesus.

The old woman: That might be.
And then again it mightn't.
That one there is married
to a dead man, and there ain't
but mighty little future in that.

The young woman *(lowering her voice as Henry passes and Hannah approaches):*

I declare, I'd let those men
carry their own water, if it was me.

The old woman:

Maybe that ain't water
to drink. Maybe that's water
to fish in.

The old man: Now Marthy!

Hannah: Good afternoon.

All three: Evening.

Hannah: Picking beans?

The old woman: What the bugs
has left us.

Hannah passes on by.

I said to myself
when Mat and them's boy got killed
in the war, and she was left
to have and raise that baby
by herself, I said it's a pity
they don't think beforehand,
before they go to marrying
and begetting and visiting
the Lord God only knows what
on some poor innocent thing
that never caused none of it.

Hannah: Like a baking or a pregnancy
the time has come to fullness
and can be no fuller.
It can't go on being
what it is. I haven't tried
to change it, but I feel
it changing. I feel it
in the air, hovering over me
and all I'm part of,
like the closeness before rain.

Henry: Hannah, come look!

He examines what he has picked up from the path, then puts it in his pocket, and again searches the ground.

Hannah: I'm coming.

Old Jack: There's a cloud in the west
and with it so hot and still
it ought to rain. I'm dry,
an old man with his death
coming. The thought of a grave
with the grass green on it
and the rain wetting it
seems bearable to me now.
Seems good. The only youth
I'll have again is in it.

The old man: Evening, Jack!

Old Jack: Evening
to you!

The old man: Mighty hot, I'd say,
for two old birds like us
to be out in the sun, stirring.
We ought to be in the shade
somewhere, setting and resting.

Old Jack: Setting and resting, hell!
Setting and resting 'll soon
make a dead man out of you.
A man like myself, with only
twenty or thirty more years
to live, has got to be *rattling*.

The old woman:
Huh! Well, it's a coming.
And you'd better be right
with your maker when it comes.

Old Jack: Well, a man has to hope
he'll somehow be overlooked.

The old woman:
Jack Beechum, it don't make
no difference how much
land you got nor how much
money you got nor who you are,
death's a coming to you.
It don't miss a one.

Old Jack: Death's very democratic.
I've heard it. Maybe it is.

The old man *(indulgently correcting):*
Jack, Marthy ain't no democrat.

Old Jack is already going on.

The young woman:
Poor old feller. Look at him.
He *ain't* got long to be around.
He can't hardly make it.

Old Jack: Piss on them! They'd like
to cover up a hard fact
by dripping pity on it.
Molasses on bad bread.
They think a man is bound
to die mourning, clamped
to the world like a trap.
They'd have me quaver and bleat,
leaving this place,
as though I didn't know
what it's cost me. They think
I'm childish, or a child.
I'm an old man, and I
know where I've been.
The churchly ones are worst.
They'd have me glad to go.
The world's cursed, they say,
and to trade it off

for Heaven and Heaven's grace
is an everlasting bargain.
It's a gamble, I say,
and for better and worse
I like this place.
The world's curse is a man
who'd rather be someplace else.

He stops to rest, and up ahead, noticing, Hannah stops also.

The walking tires me out.
I'll have to quit this.
But there's the field there
and the men at work. I want
to know what they have to say
and how the crop is.

Henry *(running back down the path to Hannah):*
I found an arrowhead! I never
found one before. Look at it.
I found it right in the path
where people wore away the dirt
over it, walking on it.
Hannah, look!

Hannah: I'm looking.

Henry: Well, why don't you like it?

Hannah: I do. I do. It's nice.

Henry: How do you reckon it got there?
Do you reckon he shot at something
and missed?

Hannah: Maybe he lost it.

Henry: Maybe he killed something.
Do you reckon? Maybe he killed
some*body* with it. In a war.

Hannah: I hope not.

Henry: Why?

Hannah: I want
things to live. I want things
that have lived to have lived,
and things that are living
to stay alive. And I hope
things that might live *will* live.

Henry *(to Old Jack, who has rested and is coming on again):*
Look, Uncle Jack, what I found.
An arrowhead. How do you
reckon it got here?

Old Jack: Ay Lord, honey,
this ground's a lot older
than anything I know.

They go on together a little way, and Hannah stops at a walled spring and fills her bucket. She offers the dipper to Old Jack.

Hannah: Drink. While it's fresh.

Old Jack *(drinks and puts the dipper back in the bucket):*
That's good. That spring never
has gone dry in my time,
though I've seen it dwindle
mighty small once or twice.
I stopped and drank here
when I was a boy, younger
than this boy, and my daddy
before me stopped and drank
here, and his daddy before him.

Moved by his thoughts, he turns away from them and goes on ahead by himself.

A long time back that spring
was flowing, cool in the summer,
in winter too warm to freeze,
pooled still and clear
where the water catches and brims
on the rock. While we've worked
and taken pleasure and suffered
and died here, it has flowed
like the sound and the feel
and the taste of what this ground

has been to us—kinder to us,
mostly, than we've been to it.
It has been the turning toward us
of the womankindness of the earth.

SCENE 3

The procession arrives at the field's edge, Hannah first with the water bucket, and then Old Jack and Henry. Burley and Jarrat Coulter and Jarrat's son Nathan are with Mat, hoeing the tobacco. They reach the row ends as Hannah comes in sight, and Jarrat draws a file from his pocket and begins sharpening his hoe. As the scene goes on the file is passed from one to another until all four have used it.

Burley: Lord, Lord, look a yonder.
Here comes that water,
and me so dry
I could spit a ball of cotton
from here to town.

He moves into the shade of a tree a little way up the fence, taking his hat off and fanning himself with it as he goes.

And hot! Lord, and the sun
done stopped dead still
up there. I *swear* it ain't
moved an inch the last hour.

The other men follow him into the shade. It is clear from the way they move that the coming of the water is an established occasion for rest.

Wilted like a picked rose!
That's me. Give a dying man
a little drop to drink, mam.

Enjoying Burley's gab, which is clearly of a sort she has heard from him before and which is at least partly for her benefit, Hannah brings him the bucket and he lifts the dipper and drinks.

My *my!* That's mighty fine!
I'm glad I lived to drink that.
When the good Lord made
spring water He surely did
have a poor hot tired man
in mind. When He made
spring water He was doing
pret near as good as He did
when He made hot biscuits
and ham and roasting ears
and blackberry cobbler and iced tea.

He looks suddenly at Old Jack, gesturing broadly with the dipper.

"Oh that one would give me a drink
of the water of the well of Beth-lehem,
which is by the gate!"
How bout it, uncle?
Am I telling it right?

Old Jack: Mighty right!

Burley: And Mat,
my friend, am I telling it
right?

Mat: You're telling it right.

Burley (*turns, arms outstretched, hat in one hand, dipper in the other*):

Nathan? My boy? Ain't I
telling it right?

Nathan: Right!

Burley: And Jarrat there, a man
who is serious about work,
and don't believe in speeches
about anything good, he
knows I'm telling it right.
Yes!

He puts the dipper back in the bucket.

So pass on, mam,
and water these folks.

Henry is showing Mat and Jarrat his arrowhead, and Burley goes over to look. Hannah passes among them with the water and they drink, thanking her. She comes last to Nathan, who has remained to the side. He drinks and puts the dipper back. The talk that follows is level and open. They are a man and woman fully experienced, and are neither shy nor coy.

Nathan: I'd like to talk to you.

Hannah: What about?

Nathan: About what
we might mean or be
to each other. Or what we
already are to each other,
I ought to say. All it needs
is talking about. You know it.

Hannah: I do. But setting free
what is from what has been
is another thing. Maybe
it can't be done now,
I've been bound so long
to what's gone by. To live
now, to really live
and hope and take pleasure
for myself, I'd almost have
to come back from the dead.
You ought not to stir things up
that are better left alone.
You ought to let us be
at peace. To get free
of the past I'd have
to bear it all again.
One time was enough.

Nathan: It's already happening to you
or you wouldn't have said that.

You ought to let me come
to see you, take you places.

Hannah: You mean let you call for me
like a beau, and take me out
to supper or a show or someplace
like that. No. Not there
in that house. You understand,
don't you? Somehow I can't
stand the thought of that.

Nathan: Maybe I understand. Maybe I think
you can stand more than you think.
Well, meet me yonder on the hill.

Hannah: When?

Nathan: After supper. Tonight.

Hannah: I won't say I will. I don't . . .

Nathan grins at her.

I know what you're thinking.
I haven't said I won't. I don't
know what to think, much less say.
When I asked you when just now
things changed. Wait. We'll see.

Nathan: What's he got?

Hannah: Henry?
He found an arrowhead.
You'd think he found a whole
Indian tribe with lodges
and fires and cornfields.

Nathan: An arrowhead is pretty exciting
for a boy to find. I used
to find them myself when I
was a boy. And they did
almost make me see
the Indians who used to hunt
and raise their crops here.
Boys find arrowheads. Men
are usually looking for something else.

Hannah: I don't like it. I'm sure
it's silly not to, but I don't.
It's a weapon, what men
seem always to have been best at.
I want to let things live.

Nathan: By that very wish, Hannah,
you're in for all the fierceness
and violence life has.
The goodness too. But you can't
escape life by loving it.

He laughs as if both commending and disparaging the neatness of his point. Hannah stands silent.

Burley: Uncle Jack and Mat,
you remember when French Chin
worked for Pap, when Jarrat
and me was boys, and how
tongue-tied he was. French said
he got awful drunk one night
and Pap worked him hard
all day the next day, and old French
said his head was hurting
and it felt so big and heavy
that every time he leaned over
it nearly pulled him to the ground,
and he said, "The sun got about a
h-hour high, and just hu-ung there."

SCENE 4

A sloping field, the top of a high ridge, not far from Mat Feltner's farm. Below, the steeper part of the slope is heavily timbered. The soil worn thin, the field has for some time been abandoned. Now young cedars the height of a man stand on it, scattered twenty or so feet apart, and the ground between them is covered thigh-deep with the lavishly blooming flowers of early July: bee balm, black-eyed Susans, Queen Anne's lace, butterfly weed. Hannah is walking up, alone, to the top of the rise. It is dusk. By the end of the scene it will have darkened until there will be no light left at all.

Hannah: If I didn't want to go
I shouldn't have asked him when.
I knew that, yet I asked.
Uncle Jack would say I'm going
because I want to. I am.
And I'm afraid to go. I feel
the way closing off behind me
as though in every track
my foot leaves in the path
a tree springs up, or a rock
too heavy to move. And I feel
the heavy beauty and grief
of what is past, as though
I'm about to go free of it
by some great pain of birth

and death. I remember a time
not long ago when I'd have cried,
feeling what I feel now,
and been unable to go on.
I know I'll never cry
more tears of that kind.
I'm up now to where I can see
all the town. It's quiet.
The lights are coming on.
The bats are filling the sky
full of their journeywork,
and I hear a screech owl
crying. Where I'm going
I've hardly begun to go.

She walks on a few steps in silence, and then she sees Nathan waiting among the little trees, still as one of them. She merely stops. They are perhaps ten feet apart, and to the end of the scene they don't come any nearer each other.

Nathan: I believed you'd come,
but I'm as glad as if
I'd doubted it.

Hannah: I'm here.
But be slow to understand
what I may mean by it.
I'm not sure myself, yet.

Nathan: I'm middling polite, and tired,
not apt to be in a hurry.

He laughs as he did earlier, as though he can't resist needling her and yet is afraid it will be taken more to heart than he intends. After a moment he seems to turn away from that subject altogether. He turns slightly away from her and stands looking over the field for a time before he speaks again.

I bought this old farm
last week. Did you know that?

Hannah: I heard it.

Nathan (*laughs*): So did I.
Two weeks before I even
signed the note, I heard
I'd bought it, and paid
three times its worth. That's
Port William. On the alert.
You know what they say?
"Tell a lie and stick to it
long enough, and it'll come true."
That's partly right. Anyhow,
I bought the place. And signed
the note. What do you think of it?

Hannah: Well, it looks awfully grown up
and neglected.

Nathan: You're right.
What else do you think?

Hannah *(hesitates):*

I wondered why, with Jarrat
and Burley owning land,
you'd want to buy a place.

Nathan: Because I think everybody
ought to have a little
land of his own, or ought
to belong to a little land.
It's part of his manhood,
not to need to ask somebody
where to hang his shovel
or his hat. And sitting around
waiting to inherit a place
is a little bit buzzardish,
wouldn't you say?

Hannah: Yes.
But is this what you wanted?

Nathan: It's not what I wanted.
It's what was here to be had,
and what I could afford.
It's my fate, you could say.

He looks away a moment longer, and then turns back to her.

Look at it. You'd hardly
believe this was virgin ground
not so long ago, and poplars
and white oaks and sugar maples

thicker than two barrels
stood on it, and the soil
was deep and black under them.
That's gone, and here are rocks
and the cedars and the weeds
that love poor land. A place
like this one 'll be hard
to live on and take care of,
more work than a better place,
and you know why?
Because for all those years
the people who ought to have
cared for it and done the work
didn't. Do you see
what I mean by fate?
There's a life here for a man
and a woman and family too,
but not as much as there was once.
And a lifetime won't be enough
to bring it back. A man
would have to live maybe
five hundred years
to make it good again
—or learn something of the cost
of not making it good.
But hard as it is, I accept
this fate. I even like it
a little—the idea of making
my lifetime one of the several
it will take to bring back
the possibilities to this place

that used to be here.
And for several evenings now
I've been coming here
and standing and looking,
and I can imagine my life
being lived here—even
the little details of it,
workdays and fencerows and such.
I could say a lot more.

Hannah: I think what you say
is good.

Nathan: Where'd you leave
the baby?

Hannah: With her grandmother,
waiting on the front porch
to see if an owl would hoot.

Nathan: Where do they think you went?

Hannah: For a walk to get cool.

Nathan: You didn't tell them where.

Hannah: Not who I was going to see.

Nathan: You're really on the sly, then.

Hannah: Please, Nathan, don't shame me.

Her voice has become suddenly resonant with feeling, and for a moment they are both quiet, acknowledging this. They seem to realize that, having spoken so, she now stands more open to him. In what she said, in the tone of her voice, their history has begun.

Nathan: No. I won't. I didn't
mean to.

Hannah: Virgil
was their son. They made me
their daughter. They've left
no kindness undone. I have
to think of them.

Nathan: But Hannah,
don't you think they want
you to live your life? I know
they do. I know Mat Feltner,
anyhow, and I know he'd never
put himself ahead of you.

Hannah: No. Neither of them would.
But they hold to what they've
lost, the way we all do.
And I can't turn away from them
as if it didn't matter
or was easy. It does matter,
and it's not easy.

Nathan: It's you.
It's not them. And it's not
what you're turning away from
that worries you. It's what
you're turning toward, all
the unknown you'll face
as soon as you turn toward me.

Hannah: When I turn toward you
it's like the world turning
away from the sun. I only
know what I knew
when I began to turn.

Nathan: And so the question is
is that enough.

Hannah: There's no
chance ever to be touched
in the light. There's only
the dark to be touched in.

Nathan: I know it's an awful risk
I'm asking you to take. I ask,
hoping just that you'll want
to take it, that it'll seem right
to you now—not last week
or next, or in twenty years.
It's not something you'll be able
to figure about, or foresee
any sure happiness in. Clear

as my hope is, I don't know
what's going to happen. The worst
you can think of, maybe.

Hannah: Once,
the worst I could think of
did happen, so I don't doubt
it can. And I can think
of worse things now. I even
fear the worst may be bound
to happen, some destruction
that men will do in crazy hope
or anger or pride—that no man
can either believe or bear.

Nathan: I know. The war has to be
on your mind, bearing proof
of the possibility of wars to come.
The ones of us who went to it
and you who lost by it
will never go free of it.
We can't see to the ground
but by looking through horror.
But fear of the worst may be
the cost of imagining the best.
I think that some, maybe
only a few, a man and woman
here and there, must be willing
to bear the cost of the worst
they foresee, and worse than that,
to allow the best a chance.

They must find the joy to do that,
to be together and live,
or the present's darker than the future.

For maybe half a minute neither says anything. It is full dark now, and there is no sound except for the night insects. And then Nathan says:

Hannah, are you here or gone?
Where are you?

Hannah: Here!

IV

A LETTER

to Ed McClanahan and Gurney Norman in California

That was a lovely time we had out there,
those months of talk and laughter, correcting us.
Our words took on a generosity of time, passing
in the free equality of men who knew each other
as boys. We escaped all deadly official boundaries
into the natural brotherhood of countrymen,
Kentucky speaking in us, mountain and river and ridge,
before a California hearth-fire, half the night.

*

Now back in Kentucky, far from you again,
I often think of those days and nights, and long
for their music and their mirth. And then
I remind myself: The past is gone. Remember it.

*

Returning, I always put on a new body,
waking in wet dawn and going to work.
Weary at nightfall, I learn again
the trusting departure into sleep, so deeply
here I might as well be gone. Already
a new garden has fallen from my hands
into the ground. Having trusted seed
to the world, how should I not be a new man?

*

The cities have forgot the earth,
and they will rot at heart
till they remember it again.
In the streets, abstraction
contends with outcry,
hungering for men's flesh.
In the city I measured time
by the life of no living thing,
but by the running down
of engines. I grew a skin
that did not know the sun. Now
once more I have shrugged
in my city skin and sloughed it off
and emerged, new-waked.

*

The streets of the broken city
bring in the vogue of the revolutionary
—another kind of politician, another
slogan-sayer, ready to level the world
with a little truth. Those who wait
to change until a crowd agrees
with their opinions will never change.

*

The man of the earth abides in the flow.
The ground moves beneath him, and he knows
it moves. His house is his vessel, afloat
only for a while. He moves, willing,
through a thousand phases of the sun,
changing as the day changes, and the year.

His mind is like the dirt, lightened
by bloom, weighted by rain.

*

The fragment of the earth
that is now me is only on its way
through me. It is on its way
from having been a tree,
a school of fish, a terrapin,
a flock of birds. It will pass
through all those forms again.

*

(for Chloe, this one)

I come into the community of the creatures:
lily and fern, thrush and sycamore,
they turn to the light, and to the earth again.
Light and leaf, man and wife,
bird and tree—each one
a blind dancer, whose partner sees.

*

And friend and friend,
together though only in thought,
our bond is speech
grown out of native ground
and laughter grown out of speech,
surpassing all ends.

*

In spring I always return
to a blue flower of the woods,
rising out of the dead
leaves whose life it is. As I look
it wears my face's shadow.
A man always overshadows
what he sees, his presence
belonging to its mystery.
So all his ideas fall short.
Unless his speech humbles him,
keeping him steadfast in love
beyond his understanding,
he goes blind to the season.
Speech can never fathom
this flower's silence. Enough
to honor it, and to live
in my place beside it. I know
it holds in its throat a sweet
brief moisture of welcome.

Early May, 1969

MEDITATION IN THE SPRING RAIN

In the April rain I climbed up to drink
of the live water leaping off the hill,
white over the rocks. Where the mossy root
of a sycamore cups the flow, I drank
and saw the branches feathered with green.
The thickets, I said, send up their praise
at dawn. Was that what I meant—I meant
my words to have the heft and grace, the flight
and weight of the very hill, its life
rising—or was it some old exultation
that abides with me? We'll not soon escape
the faith of our fathers—no more than
crazy old Mrs. Gaines, whom my grandmother
remembers standing balanced eighty years ago
atop a fence in Port Royal, Kentucky,
singing: "One Lord, one Faith, and one
Cornbread." They had a cage built for her
in a room, "nearly as big as the room, not
cramped up," and when she grew wild
they kept her there. But mostly she went free
in the town, and they allowed the children
to go for walks with her. She strayed once
beyond where they thought she went, was lost
to them, "and they had an awful time
finding her." For her, to be free
was only to be lost. What is it about her
that draws me on, so that my mind becomes a child
to follow after her? An old woman

when my grandmother was a girl, she must have seen
the virgin forest standing here, the amplitude
of our beginning, of which no speech
remains. Out of the town's lost history,
buried in minds long buried, she has come,
brought back by a memory near death. I see her
in her dusky clothes, hair uncombed, the children
following. I see her wandering, muttering
to herself as her way was, among these hills
half a century before my birth, in the silence
of such speech as I know. Dawn and twilight
and dawn again trembling in the leaves
over her, she tramped the ravelling verges
of her time. It was a shadowy country
that she knew, holding a darkness that was past
and a darkness to come. The fleeting lights
tattered her churchly speech to mad song.
When her poor wandering head broke the confines
of all any of them knew, they put her in a cage.
But I am glad to know it was a commodious cage,
not cramped up. And I am glad to know
that other times the town left her free
to be as she was in it, and to go her way.
May it abide a poet with as much grace!
For I too am perhaps a little mad,
standing here wet in the drizzle, listening
to the clashing syllables of the water. Surely
there is a great Word being put together here.
I begin to hear it gather in the opening
of the flowers and the leafing-out of the trees,
in the growth of bird nests in the crotches

of the branches, in the settling of the dead
leaves into the ground, in the whittling
of beetle and grub, in my thoughts
moving in the hill's flesh. Coming here,
I crossed a place where a stream flows
underground, and the sounds of the hidden water
and the water come to light braided in my ear.
I think the maker is here, creating his hill
as it will be, out of what it was.
The thickets, I say, send up their praise
at dawn! One Lord, one Faith, and one Cornbread
forever! But hush. Wait. Be as still
as the dead and the unborn in whose silence
that old one walked, muttering and singing,
followed by the children.

For a time there
I turned away from the words I knew, and was lost.
For a time I was lost and free, speechless
in the multitudinous assembling of his Word.

A FAILURE

They are gone, the wild
lilies that stood here in the years
past. For the loss of meeting them
again, I am less. Will they return
next year? Will I? I needed
to find them here, unfailing,
in balanced, tensed, mottled, airy,
proud vibrance of ash.

THE GRANDMOTHER

Better born than married, misled,
in the heavy summers of the river bottom
and the long winters cut off by snow
she would crave gentle dainty things,
"a pretty little cookie or a cup of tea,"
but spent her days over a wood stove
cooking cornbread, kettles of jowl and beans
for the heavy, hungry, hard-handed
men she had married and mothered, bent
past unbending by her days of labor
that love had led her to. They had to break her
before she would lie down in her coffin.

THE ILLUMINATION OF THE KENTUCKY MOUNTAIN CRAFTSMAN

Alone, he has come to the end
of the handing down of his art,
the time having little use
for such skill as his, his land
seeded with lies and scars.
So much has he suffered
in his flesh that the end of time,
the signs being fulfilled,
the unsealing of the seals,
seems only to be borne
as he has borne the rest.
On the mountain top, stunning
him like the glance of God,
the lightning struck him. Entering
at the big tendons of his wrists,
it has stayed in his body
so that the insects no longer
bite him, and in the night
he is not afraid any more.

THE WAGES OF HISTORY

Men's negligence and their
fatuous ignorance and abuse
have made a hardship of this earth.
Living on these plundered
hillsides of Kentucky is harder
for crops and for men too
than on the terraced slopes
of Tuscany or Japan, where care
has had a history centuries
old. As if chance and death
and sorrow were not enough,
we must contend with stones
laid bare by the dream of
ease to be found in money, as if
our forefathers dug in the dark
virgin loam for gold, and found
only the bare stones and the grave's
ease. Doomed, bound and doomed
to the repair of history or to death,
we must cover over the stones
with soil for tomorrow's bread
while the present eludes us.
For generations to come we will not
know the decency and the poised ease
of living any day for that day's sake,
or be graceful here like the wild
flowers blooming in the fields,
but must live drawn out and nearly

broken between past and future
because of our history's wages,
bad work left behind us,
demanding to be done again.

THE HERON

The world as men have made it is an ungainly
hardship that comes of forgetting
there is other life than men have made.
While the summer's growth kept me
anxious in planted rows, I forgot the river
where it flowed, faithful to its way,
beneath the slope where my household
has taken its laborious stand.
I could not reach it even in dreams.
But one morning at the summer's end
I remember it again, as though its being
lifts into mind in undeniable flood,
and I carry my boat down through the fog,
over the rocks, and set out.
I go easy and silent, and the warblers
appear among the leaves of the willows,
their flight like gold thread
quick in the live tapestry of the leaves.
And I go on until I see crouched
on a dead branch sticking out of the water
a heron—so still that I believe
he is a bit of drift hung dead above the water.
And then I see the articulation of feather
and living form, a brilliance I receive
beyond my power to make, as he
receives in his great patience
the river's providence. And then I see
that I am seen, admitted, my silence

accepted in his silence. Still as I keep,
I might be a tree for all the fear he shows.
Suddenly I know I have passed across
to a shore where I do not live.

SEPTEMBER 2, 1969

In the evening there were flocks of nighthawks
passing southward over the vàlley. The tall
sunflowers stood, burning on their stalks
to cold seed, by the river. And high
up the birds rose into sight against the darkening
clouds. They tossed themselves among the fading
landscapes of the sky like rags, as in
abandonment to the summons their blood knew.
And in my mind, where had stood a garden
straining to the light, there grew
an acceptance of decline. Having worked,
I would sleep, my leaves all dissolved in flight.

THE FARMER, SPEAKING OF MONUMENTS

Always, on their generation's breaking wave,
men think to be immortal in the world,
as though to leap from water and stand
in air were simple for a man. But the farmer
knows no work or act of his can keep him
here. He remains in what he serves
by vanishing in it, becoming what he never was.
He will not be immortal in words.
All his sentences serve an art of the commonplace,
to open the body of a woman or a field
to take him in. His words all turn
to leaves, answering the sun with mute
quick reflections. Leaving their seed, his hands
have had a million graves, from which wonders
rose, bearing him no likeness. At summer's
height he is surrounded by green, his
doing, standing for him, awake and orderly.
In autumn, all his monuments fall.

THE SORREL FILLY

The songs of small birds fade away
into the bushes after sundown,
the air dry, sweet with goldenrod.
Beside the path, suddenly, bright asters
flare in the dusk. The aged voices
of a few crickets thread the silence.
It is a quiet I love, though my life
too often drives me through it deaf.
Busy with costs and losses, I waste
the time I have to be here—a time
blessed beyond my deserts, as I know,
if only I would keep aware. The leaves
rest in the air, perfectly still.
I would like them to rest in my mind
as still, as simply spaced. As I approach,
the sorrel filly looks up from her grazing,
poised there, light on the slope
as a young apple tree. A week ago
I took her away to sell, and failed
to get my price, and brought her home
again. Now in the quiet I stand
and look at her a long time, glad
to have recovered what is lost
in the exchange of something for money.

TO THE UNSEEABLE ANIMAL

My daughter: *"I hope there's an animal*
somewhere that nobody has ever seen.
And I hope nobody ever sees it."

Being, whose flesh dissolves
at our glance, knower
of the secret sums and measures,
you are always here,
dwelling in the oldest sycamores,
visiting the faithful springs
when they are dark and the foxes
have crept to their edges.
I have come upon pools
in streams, places overgrown
with the woods' shadow,
where I knew you had rested,
watching the little fish
hang still in the flow;
as I approached they seemed
particles of your clear mind
disappearing among the rocks.
I have waked deep in the woods
in the early morning, sure
that while I slept
your gaze passed over me.
That we do not know you
is your perfection
and our hope. The darkness
keeps us near you.